An Easy Guide to the Wiring Inside Your Walls

Quick & Basic

House Wiring

Carol Fey

P.I.G. Press
Littleton, Colorado

Disclaimer: This manual is intended as a tool for a classroom setting. Do not attempt any of the instructions in this book without the guidance of a qualified instructor.

Published by P.I.G. Press
759 E. Phillips Drive S.
Littleton, CO 80122-2873

Publisher's Cataloging-in-Publication Data
Fey, Carol.

    Quick and basic home wiring : an easy guide to the wiring inside your walls / Carol Fey.–Littleton, CO : P.I.G. Press, 2005.

        p. ; cm.
        ISBN: 0-9672564-3-7
        ISBN13: 978-0-9672564-3-6

    1. Electrical wiring, Interior—Amateur's manuals. I. Title.

TK9901 .F49   2005
621.319/24—dc22   2005   928831

Book production and coordination by Jenkins Group, Inc.
www.bookpublishing.com

Printed in the United States of America
09 08 07 06 05 • 5 4 3 2 1

Dedication

Dedicated to my dad and granddad,
Laurence and Clarence Kersey, electricians
who brought the first wiring to many
houses and farms in Ohio.

Contents

Acknowledgments

This book would not have been possible without
the electricity wise men: Dale Watterson, Anthony Foti,
Bruce Wagner, and Patrick Fitzgerald.

I also thank these folks for their ideas and suggestions:
Paul Armstrong, Barry Engleman, Dan Holohan, Alan Levi,
Roy Weaver, Susan Jenkins, Dave Stroman, and Russ Herre.

About the Author

Carol Fey is a degreed technical trainer who has worked in the controls industry for many years. She has been honored as National Technical Trainer of the Year by the National Society for Training and Development.

Other Books by the Author

Quick & Basic Electricity: A Contractor's Guide to HVAC Circuits, Controls, and Wiring Diagrams

Quick & Basic Hydronic Controls: A Contractor's Easy Guide to Hydronic Controls, Wiring, and Wiring Diagrams

Quick & Basic Troubleshooting: A Contractor's Easy Guide to Fixing HVAC Wiring and Circuits

Introduction

My father was an electrician. My grandfather was an electrician. And I was a girl. Even though I'm a card-carrying member of the International Brotherhood of Electrical Workers (IBEW), I had to learn electricity like most folks do—by myself and by the seat of my pants.

I've found that practical electricity is a lot easier than the experts let on. There's no need to know electron theory or Ohm's law to understand house wiring.

We ordinary folks need to know things that the experts don't even think about, such as:

- A circuit is always made up of a power supply, a switch, and a load. Period. There can be more than one switch or load, but there's always at least one.

- The cord from your lamp to the wall outlet has two wires in it. One brings hot electricity into the lamp, and the other takes the used electricity out.

Information like this is second nature to some folks, but it was a mystery to me. And I don't think I'm the only one.

Other things are also confusing:

- Why does a two-wire cable have three wires, and a three-wire cable have four wires?

- When there are switches at the top and bottom of your stairway controlling one light, why are they called three-way switches when there are only two of them? And if you have three switches controlling your kitchen ceiling light, why are they called four-way?

There are a few things the experts tell us that are just plain wrong. The classic is, "Electricity is just like water—what part of that don't you understand (dummy)?" The part we don't understand is when the analogy doesn't work. True, a switch turns the flow of electricity on and off like a faucet does water. But then it gets reversed—an open switch is "off" while an open faucet is "on"!

Many people fear electricity. There's good reason to treat it with complete respect. But as with fire or a big, mean dog, thinking about it and understanding it can keep us out of trouble.

This book alone does not prepare you to fix your own wiring. The purpose of this book is to help you understand how electricity works in a house or any small building. If something electrical needs attention, keep yourself and the people around you safe—call an electrician.

Should you want to do an electrical wiring project yourself, first get some professional instruction, or use one of the many home wiring books available in bookstores, online or free at your local public library. Although there are many well-intentioned people eager to give you electrical advice, be cautious. Regardless the topic, not everyone with an opinion knows what he's doing.

Electricity Goes in Circuits

Electricity and electrical wiring always go in circuits. Circuits are made up of these three things:

- a power supply

- a switch

- a load

There can be more than one switch or load. But there's never anything other than these three things.

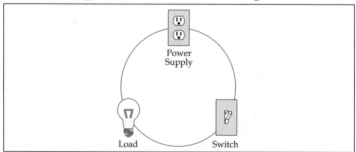

Figure 1. The basic electrical circuit.

A load changes electricity into another form of energy. A light bulb is a load. It changes electricity into light and heat. A motor changes electricity into motion. A doorbell changes electricity into sound. An electromagnet changes electricity into magnetism.

Each of these loads has a switch associated with it to turn the electricity on and off. And each has a power supply to provide the electricity.

A simple switch turns the electricity on and off, like a drawbridge controls traffic crossing a river. A switch can be a wall switch for a permanently installed light fixture. Or a switch can control a lamp, an appliance, or a wall outlet. Flip it on, flip it off—that's a switch.

A *power supply* is where the electricity comes from. We can think of a wall outlet or receptacle as a power supply. Of course, the power supply to the wall outlet is the power plant where electricity is made.

Electricity in a circuit is like a bug on a rope

Electricity has to travel in a circuit, or on a path that ends up back where it started. Electricity travels on a wire like a bug on a rope. For the circuit to work, the electricity must be able to get all the way around the circuit and back to where it started—without ever getting off.

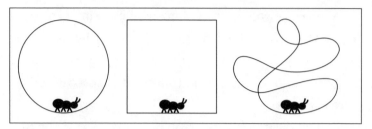

Figure 2. Electricity moves on a wire like a bug on a rope.

It doesn't matter whether the circuit is round, square, or looks like a pile of spaghetti, as long as the bug, or electricity, can get back to where it started.

Another way of looking at a circuit is that when you bring electricity in, you also have to give it a way to get out. It's a little like bringing water into a sink—you need a drain to get it out. Electricity that comes from the power plant into your house must have a way to get back to the power plant.

The wiring in a house may look like the drawing in Figure 3. But each of these "wires" is actually a cable that contains at least two wires. One wire brings the electricity into the light, outlet, or switch. The other wire takes the electricity away.

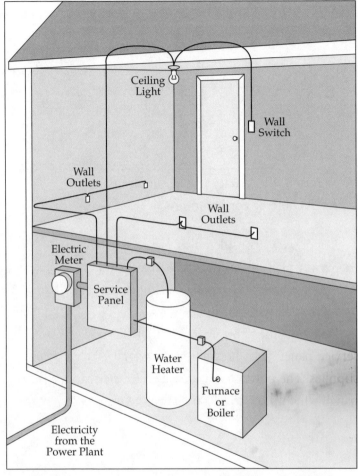

Figure 3. The many circuits in a house or other small building are shown as cables, not as individual wires. If you could see inside the cables and cords, the wires would look more like the drawing below.

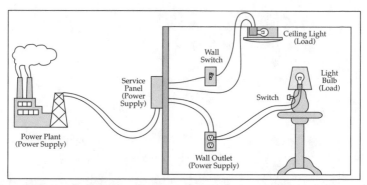

Figure 4. Electricity travels in circuits.
Each cable or cord contains at least two wires.

Circuits vary considerably in their size. The circuit from the power plant to your house is many miles long. The circuit from your service panel to a lamp is no larger than your house. But all wiring is in circuits.

Notice in the drawing above that the power plant, the service panel, and the wall outlet are all labeled as power supplies. The power plant provides electricity to the service panel, which provides electricity to the wall outlet.

Power Supplies, Loads, and Switches

The power plant—the ultimate power supply

Your electricity is made at a power plant. It comes through wires across many miles. That's not a problem, because it comes from the power plant really big and really fast.

When electricity gets to your neighborhood, it's "stepped down" at a "substation." A little closer to your house or building, it's stepped down again by a transformer. In older neighborhoods, where electric wires are overhead, this transformer looks like a barrel high on a pole along the street. In newer neighborhoods, electric distribution is underground.

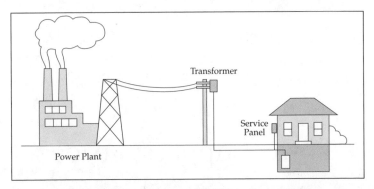

Figure 5. Electricity comes from a power plant.

From the power plant and transformers, electricity goes through either overhead or underground wires to your house or building. There it goes through a "service head" and then an electric meter so that the utility knows how much you use.

Then it goes to your "service panel." The service panel sends electricity through large-capacity breakers or big fuses to the various circuits in the building. The flow of electricity in each circuit is controlled by a safety switch. It's either a circuit breaker or a fuse, depending upon the age of the wiring.

Hot and neutral electricity

Electricity leaves the power plant packed full of energy. That electricity is called hot. The wire that carries it is called the hot wire or the hot leg. This is the electricity that can do work or damage.

When it gets to its destination, the energy in the hot

electricity is changed. It's changed into light by a lamp, into heat by a heater, or into motion by a motor. We can think of electricity as coming from the power plant on an "electricity train." Electricity comes from the power plant "hot." It's packed full of energy and looking for some place to spend it. It's like the legendary cowboy with his money on payday. If he doesn't find something constructive to do with it, it'll tear things up.

After the energy is used, there's still electricity. But this electricity is tired. It's called neutral. The wire that carries it is called the neutral leg. The electricity train uses the neutral leg to take used electricity into the earth through a grounding rod. The grounding rod extends far into the earth, where electricity is safely absorbed into the ground. That's why it's called "earth ground."

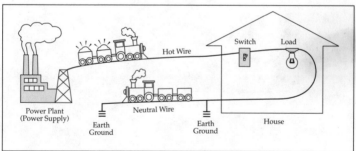

Figure 6. The electricity train.

Polarity

Have you noticed that on most plugs, one prong is bigger than the other? Most outlets have one slot bigger than the other, too. You can put the plug into the outlet only one way. The short slot is designated for hot electricity. The short plug prong leads to the "hot" side of the load. The long slot is "neutral." It goes back to earth ground. Having to match up plug prongs with the slots keeps electricity going the right direction and helps to keep us safe. Of course, don't count on it without testing. An explanation of testing comes later in this book.

The difference between "hot," usually the black wire, and "neutral," usually the white wire, is called polarity. Polarity is important because it keeps people from getting shocked. It's interesting to know that outlets and plugs weren't polarized until the 1920s.

Figure 7. A polarized plug fits only one way into a polarized outlet.

Is a wall outlet a power supply, a switch, or a load?

The answer depends upon how you're looking at it. Let's look at it from the perspective of the service panel. The panel is the power supply, the circuit breaker or fuse is the switch, and the outlet is simply a connection point waiting for a load (lamp or appliance) to be plugged into it. So in this case, let's think of the outlet as a load. No electricity goes through the outlet until a load is plugged in. In fact, the very early outlets were manually screwed into a wall fixture that could otherwise hold a light bulb.

Figure 8. Outlets in the early 1900s screwed into a light fixture socket.

If your perspective is from the outlet and whatever you plug into it, you can think of the outlet as the power supply. The circuit for a plug-in lamp is: outlet = power supply, light bulb = load, lamp switch = switch.

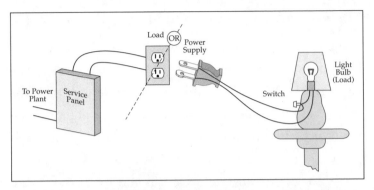

Figure 9. Depending upon your perspective, the outlet represents either a load or a power supply.

Switches

Think how bothersome a light or vacuum cleaner would be if there were no way to turn them off! Wall switches and switches built into lamps and appliances make it easy to control the device.

A switch turns the flow of electricity on and off in the circuit. Think of a simple switch as a drawbridge on the electrical circuit. When the drawbridge is open, there's no path for the electricity to pass on, so the flow stops. There is no energy for the load, which is the light or appliance. The load is "off." When the drawbridge is closed, there is a path for the electricity to go on, and the energy flows to the load. The load is "on."

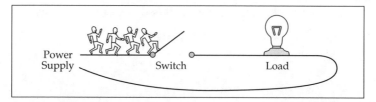

Figure 10. A switch is like a drawbridge.

Here are a few ideas that make switches easy to under-stand.

- On versus off is simple. Yet it may seem backwards to some of us that a switch is "on" when it is closed and "off" when it is open. Again, a switch is like a drawbridge. It is not like a door, gate, faucet or valve.

- Some electricity instruction is based on the "water analogy." The principle "electricity is like water" works only up to a point. The point where it stops working is switches. Water runs through pipes. The flow is controlled by a valve or faucet. When it's open, stuff flows. When it's closed, stuff stops.

The reverse is true for a switch controlling electricity. Think of a switch as a drawbridge. When the drawbridge is closed, traffic flows. When it's open, traffic must stop.

- Committee work—switches in a row (in series). Some-times there is more than one switch in a circuit. Then both switches must be "on," or closed, for electricity to flow to the load. An example of this is when you have a lamp plugged into a switched wall outlet. Both the wall switch and the lamp switch have to be "on" for the light to come on.

The switches in the service panel

Switches are for both convenience and safety. Fuses and circuit breakers in the service panel act as switches—they make sure that electricity doesn't flow in an overloaded or shorted circuit. In this sense, every circuit in a building has at least two switches. The service panel has a switch for each circuit.

In buildings that were wired before the 1960s, the service panel is a fuse box. A fuse is a one-time, disposable switch. The amperage of the circuit is marked on the fuse box, and the fuse should match. If you overload the circuit by trying to get too much electricity out of it, or of there's a short in the circuit, the fuse "blows." That means that the thin piece of metal inside the fuse opens, and that turns the circuit off.

You can see the reason that the fuse failed. If it's blackened, there was a tiny explosion in the fuse because of a short in the circuit. If the fuse is burned through, the action was slower because of an overload in the circuit. Either way, the fuse has to be replaced for the circuit to work again. It should be replaced by the same amperage fuse as the one that burned out.

The reason that putting a penny or other piece of metal in a fuse socket works, and should never be used, is that the metal is a path for electricity. It closes the switch. It's a very bad idea, because it defeats the safety factor that's the reason for having a fuse there in the first place. The heavier metal won't break if the demand on the circuit is too great, the circuit can overheat, and there can be a fire.

Replacing a fuse with a bigger, higher amperage fuse is a bad idea for the same reason—it won't burn out when you exceed the safe capacity of the circuit. That can cause the electrical wire to heat and cause a fire.

There are usually two 60A fuses for the heavy appliance circuits. If there are "odd" things going on with the electricity, such as half of the lights being dim, or the dryer turning but no heat being produced, a simple solution may be to replace both of the 60A fuses, even though one looks all right.

More recently wired buildings have circuit breakers instead of fuses. Each circuit breaker is a switch that stays closed unless the circuit is overloaded or shorted. When there's an overload, the circuit breaker switch opens. That keeps electricity from flowing into that circuit until someone resets, or closes, the circuit breaker.

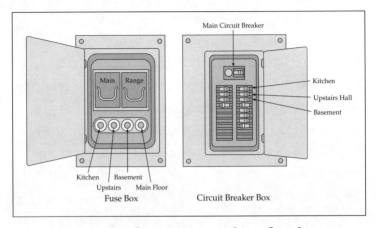

Figure 11. The service panel is a fuse box or a circuit breaker box.

Grounding and GFCI Outlets

In the 1960s, grounding began to be used in addition to polarity. A third wire was added to electrical circuits and cable. That wire is usually green or bare copper. The idea with grounding is to give stray electricity a place to go harmlessly. It goes to earth ground—the same place that the neutral wire goes.

It's important for ground and neutral to be separate wires. That's so that neutral electricity doesn't accidentally get in the grounding system and cause problems.

In electrical code, the grounding wire is sometimes called the grounding conductor. In the same terminology, the white neutral wire is called the grounded conductor. The black hot wire is called the non-grounded conductor.

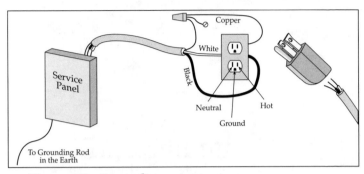

Figure 12. Grounding.

An alternate grounding method in early wiring was a "conduit." The hot and neutral wires were encased in thin steel pipe. The metal served both as a physical protection for the wires and as grounding.

To understand how grounding works, let's look at some very basic rules of electricity:

- Electricity is lazy and is always looking for the easiest way to the earth. That's usually the neutral wire.

- If electricity can't find a proper path to ground, you'll do just fine (ouch!).

When things are going right with house wiring, electricity comes in by way of the "hot" wire, goes through the load (light bulb, power tool, etc.) and goes back out through the "neutral" wire to the house grounding rod. If electricity doesn't make it back through the neutral wire, hopefully we've given it an alternate path through the grounding wire.

Back in the days when water pipes were metal, they were sometimes used as the path to ground. But plastic pipes and utility companies insulating the pipes have made that unreliable.

The grounding wire keeps us humans safer because it's usually an easier path to ground than through us. If there's no ground, though, we're fair game. The reason that standing in a puddle of water or on a metal ladder can be dangerous is that water and metal are great paths for electricity to get to ground, and we can help it along.

Grounding is needed when something goes wrong and there's stray electricity. Sometimes, electricity can hop off its prescribed circuit. Let's say that inside a lamp, the hot wire somehow touches metal. Going through the light bulb is a lot of work for electricity. If there's an opportunity, it's a lot easier for it to jump off the circuit and try an easier route. If it does and there's no grounding, when you touch the lamp—zap, you're the electrical conductor.

The same loose-wire situation could happen inside a wall switch or outlet. The loose wire touches the metal box, and the electricity is on the loose and looking for where to go next. If you make yourself available, you'll do just fine.

However, if the circuit is grounded, the electricity prefers the grounding wire over you. Good choice!

Grounding appliance ignition systems

Grounding is required for a different reason inside appliances with flame ignition systems, such as stoves or furnaces.

These require what's called a burner ground. This is a wire from the electronic brain, called an ignition module, back to the burner. Its purpose is to insure that an electrical signal gets back to the module when flame is proven before releasing any quantity of fuel.

Ungrounded wall outlets

In buildings where the electricity was put in before the 1960s, you see wall outlets with only two slots. It's possible, but unlikely, that those outlets have since been grounded. It's easy to test to see if they are grounded, and we'll talk about that in a bit.

Most folks don't understand grounding, so the situation of the older two-slot outlet and a modern three-prong appliance plug deceptively looks like a matter of making three prongs fit into two holes. It may be tempting to remove the third prong, but that's not a good idea.

A second choice is to use an adapter. A lot of us innocently think an adapter is a safe solution. It certainly allows the three-prong plug to fit in the two-slot outlet. But it can be misleading. Even though the grounding prong is in the adapter, the appliance may not be grounded at all.

Here's where the adapter can fall painfully short.

- If the outlet isn't grounded, it's still not grounded when you use an adapter.

- If the outlet is grounded, the adapter still needs to

be connected to the outlet. Just plugging it in doesn't do the job.

The chances of both of these conditions happening are slim. So while the adapter lets you plug in the appliance, grounding is often an illusion. Certainly, if the adapter isn't permanently installed on the outlet, it's not giving you any safety.

There are instructions for testing an outlet later in this book.

GFCI outlets

In an older installation where no grounding wire is available, the GFCI (ground-fault circuit interrupter) outlet is a way to add protection to the existing circuitry in place of grounding.

The GFCI outlet is now required for new installation of outlets that are within six feet of a water source. It's a specialized outlet installed in areas such as kitchens, bathrooms,

 and outdoors. The GFCI is able to detect a tiny change in amperage. Such a change could be caused by contact with water or the sink, since it's connected to water. For example, if a hair dryer fell into the sink, we'd want the electricity to the hair dryer to be shut off immediately. That's what the GFCI does.

Figure 13. A GFCI outlet.

Volts, Amps and Watts

Volts (voltage), and amps (amperage), and watts (wattage) are the measurements of electricity that you hear about most. Let's take the mystery out of them.

Voltage. Remember the electricity train? The hot electricity coming from the power plant is pressurized. The pressure is called voltage. Ordinary house wiring is standardized at 120 volts (V). This is called house current or line voltage. This is the voltage you can expect to have in your house, plus or minus 5 percent. It may not be actually 120V. If you test the voltage in an outlet, you'll find that it can vary by time of day or the season of the year. These differences don't affect the performance of lights or appliances.

Sometimes you hear of 110V or 115V. Over the years, standard voltage has moved up from 110V to 120V.

There are also line voltages higher than 120V. For example, an electric water heater or clothes dryer is standardized at 240V. Over the years it has grown from a standard 220V, to 230V, to 240V. This 240V electricity is available at the service panel. Red and black wires are usually used for 240V.

Amperage. Whereas voltage is a measurement of static pressure, amperage is a measurement of flow, or how much. A larger wire can carry more amperage, just like a larger pipe can carry more water. Amperage is also called current.

Every circuit has an amperage, or amp, rating. It's limited by the size of the wire. An ordinary 120V household circuit is rated at 15 amps (A), but can also be 20A or 30A. There are also 240V circuits, rated at 40A and 60A.

If you try to get, or "pull," too much amperage through a circuit, the fuse or circuit breaker turns the circuit off. We have that safety precaution because trying to pull too much amperage makes the circuit heat up and can cause a fire.

More amperage requires heavier wire to carry it. Wire is measured by a term called "gauge." The larger the gauge, the smaller the wire. It's the same idea as shotgun shells, where 12-gauge is larger than 20-gauge.

For 120V, 15A light fixtures and outlets, 14-gauge wire could be used. For a 240V, 40A electric range, 8-gauge wire could be used. Wiring problems are often related to too small a gauge being used, probably with the intent of saving money.

You can tell by looking at an outlet what its voltage and amperage limitations are. Here are the common ones:

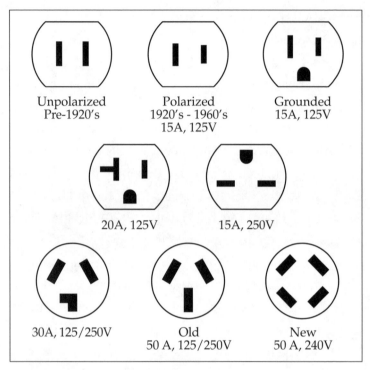

Unpolarized
Pre-1920's

Polarized
1920's - 1960's
15A, 125V

Grounded
15A, 125V

20A, 125V   15A, 250V

30A, 125/250V   Old
50 A, 125/250V   New
50 A, 240V

Figure 14. Types of outlets and their voltage and amperage limitations.

Wattage. Most loads have a wattage rating. An ordinary light bulb might 60W. A power tool might be 1500W. Wattage is simply volts multiplied by amps.

If an appliance has only an amp rating and you want to know its wattage, simply multiply the amps by the voltage (most likely 120V).

VA is also volts multiplied by amps—the same formula as for wattage. Inductive loads, such as motors and transformers, have a VA rating. Resistive loads, such as light bulbs, are rated in watts. But it's the same measurement.

Figuring circuit capacity. Let's say you're wondering if you can safely plug in a new appliance. Or you're finding that your circuit breakers or fuses keep going out. Here's what you can do to figure the circuit capacity. Look on the circuit breaker or fuse for that circuit and get the amp rating. Let's say it's 15A. Multiply that by the circuit's voltage (120V unless it's a dedicated heavy-appliance circuit). That gives you 1800W.

To give a margin of error, let's assume just 80 percent of that. Multiply 1800W by .80, and you get 1440W. Now add up the wattage of every load (fixtures, plug-in lamps, and appliances). If the total is less than 1440W, you're in business. If it's more, you need to remove loads until it's less.

Wires, Cords and Cables

Electricity that goes into a circuit has to get back out. That means that a lamp must have two wires. But when you look at a lamp, there's only one cord.

What you're seeing when you look at a lamp cord is a cable. A cable is an insulated covering with wires inside. There are two wires inside a lamp cord that has a two-prong plug. One is for bringing the electricity into the light bulb. This is the "hot," or black, wire. The other is for taking the electricity back out. It's the "neutral," or white, wire.

If you look closely at the edges of a lamp cord, you'll probably see that one is smooth and the other is ribbed. The ribbed edge is to be connected to the larger neutral prong. The smooth edge is to be connected to the smaller hot prong. If you replace a plug, paying attention to this detail insures that the lamp switch controls the hot leg. If the cord has a three-prong plug, there are three wires inside.

The third wire is for grounding.

One end of the black hot wire should be connected inside the plug to the smaller of the two prongs. The other end of the black wire should be connected to the lamp switch. The hot wire gets connected to the switch so that the switch will turn off the hot electricity.

One end of the white wire is connected to the larger, "neutral" plug prong. The other end connects to the lamp wire not going to the switch.

Notice that in the lamp we have a complete circuit of power supply, switch, and load (light bulb). The switch and load are contained in the lamp. You can think of the power supply as the outlet, the service panel, or the power plant— they're electrically the same. The three parts together make the circuit.

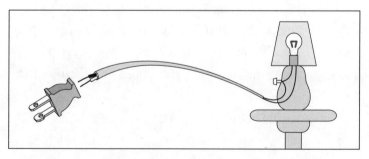

Figure 15. A lamp cord contains two wires (three if it's grounded).

Modern house wiring is now usually done with cable. From the outside, a cable looks like one fat wire. But inside, there are two or more smaller wires. You can see the separate wires on the ends of the cable. The internal wires are colored by their insulation. In a two-wire cable, one is black and one is white.

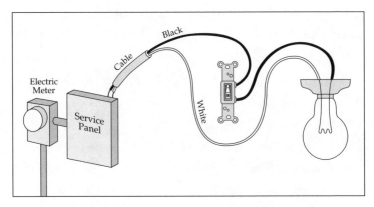

Figure 16. A pre-1960s two-wire cable contains a black hot wire and a white neutral wire.

There is some interesting history in the evolution of house wiring and cables. Of course there is overlap in the dates these were used, depending upon local practice and code. "Historic" wiring is still in use in older buildings today.

House wiring before the 1940s was done in what's called knob and tube wiring. Wires were in a cable of rubberized fabric. It was run over porcelain insulators that were mounted to the walls. Through the 1940s, flexible metal cable called Greenfield or BX was used. The metal provided external protection for the two wires it covered. The metal sheath also provided grounding. This idea was continued into the 1970s with metal conduit.

Non-metallic (NM) rubber and fabric-covered cable was used from the 1930s into the mid-1960s. As this type ages, it's likely to crack and break. Originally, it couldn't be grounded because it had neither a ground wire nor metal covering.

Beginning in mid-1960s, a grounding wire was added to NM cable and the cable was covered with vinyl. This is used today and is often called by the brand name Romex. When it contains black, white, and ground wires, it's called two-wire cable. When it contains an additional red wire, it's called three-wire cable.

Historically, both copper and aluminum wire has been used. Old aluminum wiring has gotten a bad name, mostly because of the connectors. Because aluminum expands and contracts more, the connections are more likely to work loose.

Here are some keys to safe wiring:

- It's essential that the wiring be grounded. If it's not, installing GFCI outlets is a solution.

- The cable covering must not be cracked or broken.

- The wiring must be large enough to carry the amperage of the circuit.

- The connections must be solid.

Grounding wires

The purpose of the grounding wire, and the third prong in plug, is to give electricity a safe place to go if it gets off track from the wires it's supposed to be in. There's usually no electricity in the grounding wire, so it can be bare copper.

We omit the grounding wires in many of the drawings for the sake of simplicity. That doesn't mean that it isn't important. We'll show grounding wiring separately. Of course, in real life wiring, it is all done simultaneously.

Connecting wires to each other

Wires can be connected together so that electrically, they're one continuous wire. The bare ends of the wires are twisted together. They are then covered with a screw-on cap called a wire nut.

Wires from two different cables can be connected together.

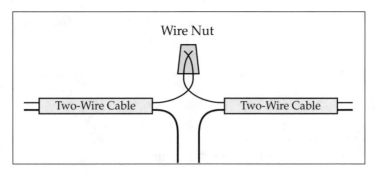

Figure 17. Wire connection from two cables.

Multiple wires can also be connected together.

Code typically requires that every connection be enclosed in an accessible metal junction box.

Series versus Parallel Circuits

The two basic patterns for wiring are called series and parallel. The two methods use exactly the same components and wires/cables, yet the results are very different.

Series means connecting everything together like beads on a string. Electricity has to go through every part to get to any other part. The classic example is old-fashioned Christmas tree lights. The well-known disadvantage is that if one light load burns out, none of the other lights works. Clearly, this isn't a good way to do house wiring.

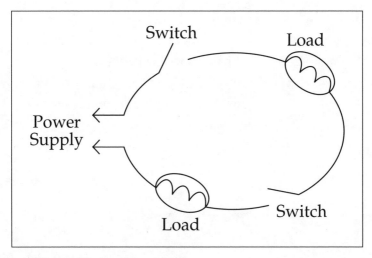

Figure 18. Series wiring.

Another disadvantage of series wiring is that you can't switch individual loads/lights. If one switch is turned off, everything in the circuit gets turned off.

Still another disadvantage is that the loads all have to share the electricity, and none gets as much as it wants.

If you have two 60W light bulbs wired in series, each gets much less electricity than it needs. They're both dim. If you have a 60W and a 25W bulb, the 25W bulb looks fully bright. The 60W bulb looks like it's not even on. But of course it must have electricity going through it, else the series circuit would not work. There is a tiny bit of light and heat coming from it.

We do put switches in series in some situations. Every wall switch or lamp or appliance switch is in series with a circuit breaker or fuse in the service panel. Both must be "on" for electricity to flow.

Parallel wiring is the primary method for house wiring. A simple definition of parallel wiring is: a number of simple circuits (power supply, switch, load) share a power supply. The following diagram shows a parallel circuit made of the same components and wires as the series circuit above.

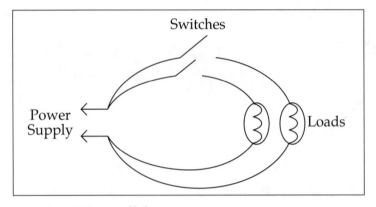

Figure 19. Parallel wiring.

Parallel wiring overcomes all of the disadvantages of series wiring:

- Each load has access to the power supply without going through the other(s). If one bulb burns out, the others are unaffected.

- Each load can be switched individually. Turning off one load's switch has no effect on the other loads.

- Each load gets all the electricity it needs. If you have two 60W light bulbs, both light fully bright.

Here's an application: Imagine the row of light bulbs that are often above a bathroom vanity. Are they wired in series or in parallel? Your clues are that they all burn fully bright, and if one burns out, the others still work. They are wired in parallel.

Of course there's a limit to how many bulbs you can have on a circuit. The limitation is that the combined amperage of the loads (bulbs and motors) must be less than the amp rating of the circuit. If it's a 15A circuit, the loads must total less than 15A. If 15A is exceeded, the circuit breaker in the service panel trips or the fuse blows.

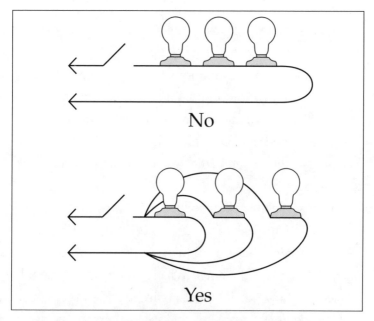

Figure 20. Vanity lights are wired in parallel.

Wall outlets are wired in parallel, not in series.

Inside the Walls

A house is wired with many different circuits. All of them start at the service panel. They are parallel circuits, with the service panel as their common power supply. Each circuit is switched by either a fuse, if it's an older house, or a circuit breaker. These switches are what's called *normally closed*. They stay closed, or "on," unless there's an unsafe situation. In that case, the switch in the unsafe circuit opens and turns the electricity off.

The fuse, or the circuit breaker, is marked for how many amps the circuit can carry: 15A, 20A, 30A. Each fuse or circuit breaker controls a circuit that provides electricity to a particular area of the house. The circuit breakers might be labeled, such as "kitchen and family room," "upstairs bath and bedrooms," "range," etc.

If they aren't labeled, you can do it yourself. Turn on everything in the house. Then turn off all the fuses or circuit breakers. Turn on just one and see which part of the house comes on. Then label the circuit in the panel accordingly. Continue with all the other circuits.

Since a circuit can serve an entire area of the house, several outlets, wall switches, and light fixtures are on the same circuit. They can operate independently because they're wired in parallel.

If you could see inside the walls, you'd see the cable, not the individual wires inside it. You wouldn't notice at first that at the same time electricity can go through the first outlet, it can also go around that outlet to everything else in the circuit.

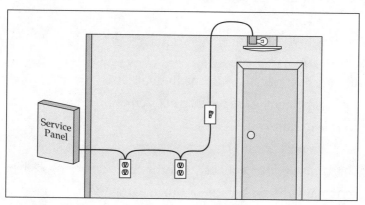

Figure 21. Wiring inside the walls.

Wiring outlets

Let's look closely at outlets wired in parallel. Let's see how electricity can simultaneously go through and around an outlet.

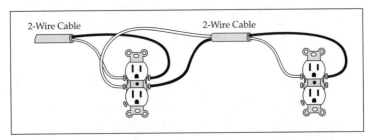

Figure 22. Electricity can flow through the outlet and, at the same time, around it. Grounding wires are omitted for drawing clarity. (Note: This drawing shows continuous wires. In modern wiring, for safety, a wire nut connection is required in the white neutral wires.)

For simplicity, we'll look first at an ungrounded outlet. When you look at an outlet without its cover, there are four terminal screws. If you actually look at one in your wall, don't touch any wires or terminal screws, because some of them can hurt you.

"Terminal" means ending. The terminal screws are where the wires end, or connect. On the left side are two silver terminal screws. These are for the neutral (white) wires. On the right are two brass screws. These are for the hot (black) wires. The screws on each side of the outlet are connected to each other by a brass strip—brass screw to brass screw and silver to silver.

A two-wire cable from the service panel at the left brings a black and a white wire to the first, or left, outlet. The white is attached to the top silver terminal. The black is attached to the top brass terminal. A second two-wire cable connects this first outlet to a second outlet.

From the second cable, the white wire connects to the bottom silver terminal of the first outlet. The black wire from the second cable connects to the first outlet's bottom brass terminal.

Finally, the wires at the other end of the second cable attach to the second outlet. The white wire goes to the top silver terminal. The black wire goes to the top brass terminal.

Electrically, two things can happen at the same time. If there's something plugged into the first outlet, electricity comes in the black hot wire, goes through the lamp or appliance that's plugged in, and leaves through the white neutral wire back to the service panel.

Whether or not there's a load plugged into the first outlet, electricity can move on to the second outlet. Electricity moves from the service panel to the top black terminal of the first outlet, and along a connector to the second terminal of the first outlet. It then moves on to the second outlet.

If there's a load plugged into the second outlet, electricity goes from the black wire through the load, through the white wire back to the bottom silver terminal of the first outlet, through a connector on to the top silver terminal of the first outlet, and back to the service panel.

Outlets are often called "duplex outlets" because they are in pairs, like duplex apartment units. The number of outlets is limited only by the amp rating of the circuit.

If the wiring was done in the 1960s or later, the cable also contains a third wire, which is a grounding wire of bare copper or insulated green. It's connected to a green screw on each outlet, or to a grounding screw in the outlet box. Here is an illustration of how the grounding wiring would look without the black and white wires:

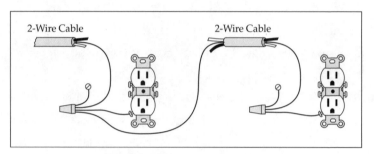

Figure 23. Cable and outlets with grounding wire. Black and white wires are omitted for drawing clarity.

And here is a drawing of the black, white, and grounding wiring all together:

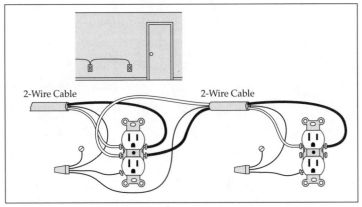

Figure 24. Black (hot), white (neutral) and copper or green (grounding) wires together.

Wiring switches and loads

A switch is connected with the same wires as an outlet, but in a different way. You've just seen how both the hot and neutral wires, and hopefully the grounding wire, are connected to an outlet. On a switch, only the hot and, hopefully, the grounding wires are connected. That's because the switch "breaks only the hot leg" of the circuit.

Remember that "breaks" means "turns off," or creates an opening like a drawbridge. The neutral leg doesn't need to be broken, or turned off, because the electricity it carries isn't energized. The neutral wire carries the used-up electricity back to where it came from.

A wall switch has three terminal screws. In the next figure, look at the pair of terminals on the right side. One is for the hot electricity to come in, and one is for it to go out. If the switch is closed (turned on), the electricity passes through the switch as if the switch weren't even there. If the switch is open (turned off), it stops the electricity.

Here's how switch wiring looks if the switch is installed between the power supply (service panel) and a ceiling light (load):

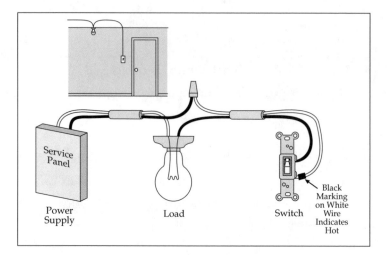

Figure 25. Wall switch in a circuit—power supply, switch, load. The switch is in the hot leg. Grounding wires are omitted for drawing clarity.

Recall that there are actually two wires in the cable. The switch is in only the black wire. The white wire goes around the switch.

There are two different wires coming to the switch. The black wire from each cable is connected to one of the terminal screws. The white wires from the two cables are connected together. They aren't connected to the switch at all. Electrically, this means that the used-up neutral electricity in the white wire goes around the switch and back to the service panel.

The switch and the load (ceiling light in this case) can be wired together in the reverse order, with the load before the switch. Electricity doesn't care which order they're in, but the wiring is different.

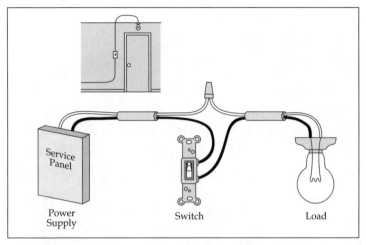

Service Panel

Power
Supply

Switch

Load

Figure 26. Wall switch and fixture circuit—power supply, load, switch. The white wire is coded black. Grounding wires are omitted for drawing clarity.

Here's what's different from the previous drawing. The switch always needs to have two black wires connected to it, and no white wires. A cable has only one black and one white wire. When the switch is "at the end of the run," through, only one black and one white wire are available.

In this situation, the electrician uses the white wire to

bring power to the switch as if it were black. This white wire is usually marked with a strip of black tape or a black mark to designate that, electrically, it's a black wire. Because white neutral wires aren't connected to a switch, assume that every wire, no matter the color, is hot if it's connected to a switch.

Grounding the switch and fixture

The grounding terminal is the third screw on a switch. It stands by itself and is bare copper or colored green. Older switches aren't grounded. In newer switches, the grounding wire is wired to the switch like it is to an outlet. If the switch box is metal, the grounding wire is connected to both the switch and the box. A "pigtail" is used to make the connection. A grounding wire is also attached to the light fixture.

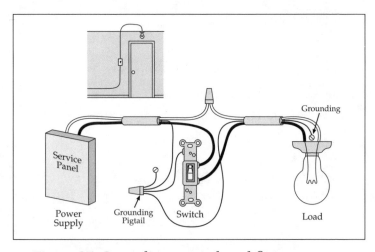

Figure 27. Grounding a switch and fixture.

Wall switch, fixture, and outlet together in a circuit

A wall switch, a fixture, and an outlet can all be in the same circuit. The fixture is switched by the wall switch. The outlet operates completely separately from the switch and fixture because they are wired in parallel.

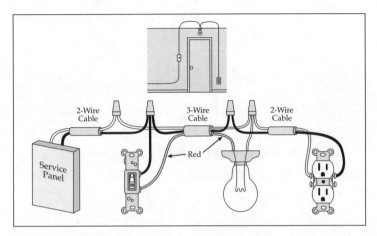

Figure 28. A circuit including a wall switch, fixture and outlet. Grounding wires are omitted for drawing clarity.

Switched outlet

Some outlets are switched. The National Electrical Code requires a switched outlet in any room that doesn't have a fixed ceiling light. You can have both a switched and a non-switched outlet in the same outlet box by removing the connector tab between the two brass terminals.

Here's the wiring for a switched outlet:

QUICK AND BASIC HOUSE WIRING

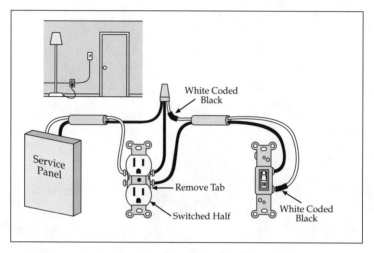

Figure 29. A switched outlet in a box with a non-switched outlet. Grounding wires are omitted for drawing clarity.

Three-Way and
Four-Way Switches

Three-way switches—two switches controlling a light

Here's the switching scenario: You need to turn on the stairway light before going upstairs to bed. Once you get upstairs, you need to turn the light off from a switch at the top of the stairs. Either switch needs to be able to turn the same light on or off. This requires special switches that aren't simple on-off.

These are called three-way switches, even though there are only two of them. If you look closely at them, you'll see that there's no on-off marking like you find on a simple switch. That's because either position can be on or off.

"Three-way" means that there are three components in the circuit—two switches and the light(s). And there are three switch terminal screws inside the switch. If there are three or more switches controlling the same light(s), they are called four-way switches. There will be more on those later. Let's first look more closely at three-way switches. The circuit has to be wired in a special way, and not the way you might think.

It might seem that two simple on-off switches and a light could be wired together one after the other (see the next figure). This would be simple series wiring. But simple switches wired in series won't turn on the light unless both switches both are turned on. What we want for the stairway is for each switch alone to be able to turn the light on or off, regardless of the other switch.

Simple switches in parallel won't work either. Wired that way, either switch could turn the light on. But to turn the light off, both switches would have to be off.

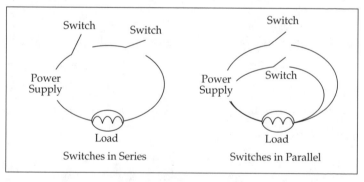

Figure 30. Simple switches wired in either series or parallel don't work like three-way switches.

Three-way switches are installed in pairs. Each switch has three terminal screws—a darker common (C) and two travelers (T). The two traveler screws are interchangeable with each other. Different manufacturers put the common and traveler screws in various positions. For example, the terminal screw patterns below are electrically the same. This common terminal is connected to the black hot leg coming from the service panel.

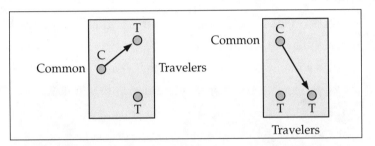

Figure 31. These different terminal screw patterns are electrically the same.

There are two possible paths for electricity to take through a two-way switch. It's like a "Y" in a road. Inside the switch, electricity goes from the single common terminal to either of the traveler terminals. It will choose the direction where the switching position is closed. Either path takes the electricity to the next switch. There it again has two possible paths through the switch.

Three-way switches can be wired in three different configurations—the two switches before the light fixture, the light fixture between the two switches, or the light fixture before the two switches.

Notice that there are three cables in each layout. A two-wire cable (black and white wires) always comes from the service panel. A three-wire cable (black, red and white) always connects the two switches. Each of the two traveler screws on the first switch may be connected to either of the traveler screws on the second switch. Depending upon the layout, the third cable may be either two-wire or three-wire.

Notice that as with loads anywhere, there is always a black (hot) and white (neutral) wire connected to it. If you see two white wires, one is *white coded black*. That's because the cable doesn't always give you the colors you need. So if you have a white, but need a black, you may designate the white as black by marking it with a piece of black tape or a black marker.

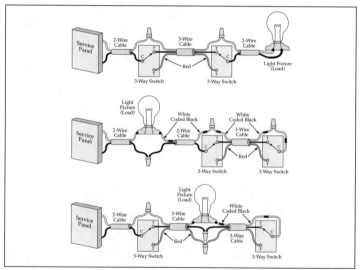

Figure 32. Three different three-way switch layouts. Grounding wires are omitted for drawing clarity.

Four-way switches— three or more switches controlling a light

Some rooms, such as a kitchen or living room, have more than two entrances. For these, we use four-way switches. Four-way means three or more switches. It's easy to move from the three-way switches that we just looked at to four-way. Just add a four-way switch between the two three-ways. In fact, you can add as many four-way switches as you need.

A four-way switch has four terminal screws. The screws are in pairs. The pairing may vary by manufacturer, so it's important to look closely at the switch for which pair together. For our purposes here, we'll pair the upper two screws with each other and the lower screws with each other.

Here's a layout for three switches and a light fixture:

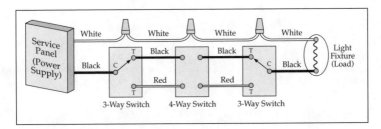

Figure 33. A four-way switch inserts between the two three-way switches. Add as many as you want. See the next drawing for a look inside the switch. Grounding wires are omitted for drawing clarity.

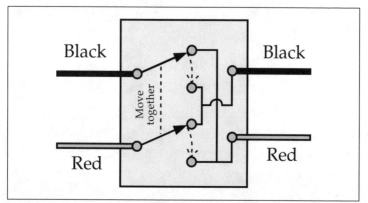

Figure 34. Inside a four-way switch.

Troubleshooting

The seemingly simple stuff that trips us up

Oh, the basics that we love to skip! When something electrical doesn't work, it's so easy to jump to the difficult solutions. Start with the easy questions: "Is it turned on?" and "Is it plugged in?" Don't move on until you absolutely have a "yes" to these.

The next question to ask is whether or not everything else in the house is working. If nothing else is, or only half is, you may have a power outage from the utility. Find out if folks around you have power. If they have power and you don't, the next place to go is your service panel—the circuit breaker box or fuse box. Listen first for a sizzling noise. If there is one, stay away and call an electrician. Otherwise, open the box and look for any circuit breaker that's "off," or a damaged or loose fuse. Turn the circuit breaker on, or replace the fuse with one of the same amperage.

Safety

If you isolate your problem to a particular outlet or switch, you can do some simple troubleshooting. Basic electrical troubleshooting is easy and safe if you pay attention to what you're doing and follow the rules. The rules are:

- Troubleshoot only when you are fresh and alert. As with using power tools, that means no intoxicants recently.

- Figure out ahead of time what you're going to do and how you're going to do it. There's a list of publications about electrical repairs at the end of this book.

- Don't touch any metal for any reason, even if you think there's no electricity there. This includes all wires, screws, and water pipes.

- Don't have water anywhere near your work.

- As with scuba diving, have someone nearby who can get you help on the very outside chance that you get yourself into trouble.

- If you're not sure of what you're doing, call an electrician.

Tools for testing

The simplest way to test an outlet is to plug in something that you know will work if there's power. Be aware that the top and bottom of a duplex outlet may not be the same, especially in a kitchen. They can have two different fuses or circuit breakers. Never assume that if one doesn't have power, the other doesn't either. Also, one may be switched and the other not.

You can do basic troubleshooting with just a few inexpensive tools that are available at hardware stores. Some of these are a neon tester, a three-prong outlet analyzer, a continuity tester, a pen-type mini-voltmeter, and a multi-meter. You may also need an ordinary screwdriver to remove a cover plate.

Testing a three-slot (with grounding slot) outlet

Figure 35. Using an outlet analyzer, test both parts of a duplex outlet separately.

By far the easiest way to test a three-slot outlet is with a three-prong outlet analyzer. You don't have to take anything apart. You simply plug the analyzer into the outlet. The combination of lights that appear tells you if the hot, neutral, and grounding wires are connected correctly, are not connected, or if they're reversed.

Testing with a neon tester, multi-meter, or pen-type voltmeter. The three-prong outlet analyzer is the easiest tester to use. But if you have a different tester, or if you don't have a three-slot outlet, here's how to test for outlet power, polarity and grounding. Make sure that you touch only the insulated part of the device's probes, never the metal tips.

If you're using a multi-meter, turn it "on" to the "volts AC" setting for the highest voltage. Volts AC may be indicated by a V with a squiggle above it. To test for power with any of these tools, put one probe in the short (hot) slot and one in the long (neutral) slot. The neon tester should glow. A voltmeter or multi-meter should give a reading around 120V. If you see nothing, there's no power at the outlet. That means that either the power is shut off at the service panel, the outlet is defective, or the switch controlling it is off.

To test for polarity, put one probe in the short (hot) slot of the outlet. Put the other probe in the grounding slot. You should see a glow, or a reading. If not, put one probe in the long (neutral) slot and the other in ground. If you get a glow or reading, that means that hot and neutral are reversed inside the outlet.

To test for grounding, the probes should be positioned as for testing polarity. If there is no glow at all, the outlet isn't grounded. That's possible even if there's a grounding slot in the outlet. Someone could have replaced an old non-grounded outlet with one that is grounding-capable, without installing a grounding wire. Or the grounding wire could be detached. On an old steel pipe conduit system, the conduit serves as the grounding, and there is no grounding wire.

If the outlet doesn't have a grounding slot, it's still

possible that it's grounded. Here's how you test for it: Put one probe in the short (hot) slot, and touch the other probe to the screw in the middle of the cover plate. Make sure there's no paint on the screw, and that the screw isn't plastic. If the tester glows, the outlet is grounded.

If the tester doesn't glow, put one probe in the long (neutral) slot, and the other on the cover screw. If the tester glows, the outlet is grounded, but the polarity is reversed. If the tester doesn't glow in either position, the outlet isn't grounded. If the outlet tests as grounded, then you can safely use a grounding adapter. But you have to permanently attach it to the outlet.

To attach the adapter to the outlet, turn off power from the outlet at the service panel by turning the circuit breaker off or removing the fuse. Test the outlet for power (above) to make sure there's no electricity. Remove or loosen the cover plate screw, plug in the adapter in the outlet, put the screw through the ring on the adapter, and replace the screw into the cover plate. Turn the power back on.

Testing a switch

You can test a switch with either a continuity tester or a multi-meter set on "continuity" or "ohms" (Ω).

You must first turn off power at the service panel. That's because either of the testers uses its own battery to send a small amount of electricity through the switch. It's harmful to the testers to have any other electricity there. Getting electricity back confirms that there is a path through the switch. That means that the switch is good.

The continuity tester is the easiest to use. Turn off power at the service panel. Test the tester itself by touching its tip to its clip. The tester lights. Now put the switch in the "on" position. Attach the clip to one of the screw terminals on the right side of the switch. Touch the tester tip to the other screw on that same side. If the switch is good, the tester lights.

You can do the same process with a multi-meter. Again, make sure you've removed power from the switch circuit. Set the meter for "continuity" if it has that setting. Otherwise, set it for ohms (Ω) symbol. When you touch the two meter probes together, the meter will make a beeping sound. Now, make sure the switch is in the "on" position. Touch a probe to each of the two switch terminals on the right side. The meter sounds if the switch is good.

If the switch doesn't test good, it needs to be replaced. Either call an electrician or consult a home wiring book for how to replace it.

Other Interesting Things to Know

Low voltage circuits

Behind the scenes in a house or building there is voltage lower than 120V house current. It's called "low voltage" and is defined as 24V or less. Heating and air conditioning systems are often controlled by a 24V circuit. A doorbell uses 24V or less. Advantages of low voltage are close temperature control for HVAC and the fact that there are no licensing requirements to install or repair low voltage wiring. Also, if it shorts out, it won't start a fire.

Low voltage wiring is done in the familiar circuit of power supply, switch, and load. All of these components have to be rated for the lower voltage, not for 120V.

There is likely a circuit from the service panel dedicated to the furnace. Air conditioning may use the same one, or another. One of the "loads" on this circuit is a low voltage transformer. The transformer changes 120V into 24V.

One-fifth of 120V is 24V. The "primary" side of the transformer, where the 120V enters, is a load because it changes electricity into another form of energy—magnetism.

The reduced-down side, or secondary side, picks up one-fifth of the magnetism and changes it into electricity. This side of the transformer is the power supply for a low voltage circuit. There's no polarity (hot or neutral) on this side of the transformer because the change from 120V to 24V was done through electromagnetism. However, for some solid state loads, it's important to observe the "load" and "common" designations on the transformer secondary. And if for some reason you decide to wire transformers together, it's essential to wire like sides together.

The switch in this circuit is your thermostat. A thermostat is just a switch that turns on and off by temperature change. The load or loads is a control, such as a valve, that opens to allow heating when the thermostat switch turns on.

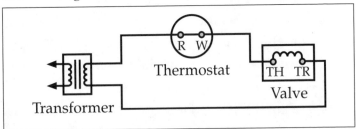

Figure 36. A low voltage control circuit.

Your doorbell circuit uses a separate transformer if it's a different voltage. In the doorbell circuit, the transformer is the power supply. The doorbell button by the door is the switch. The load is the device that makes the sound when the button is pushed.

Indoor or outdoor low voltage lighting is like any other lighting, but with a transformer supplying low voltage. The loads must be wired in parallel, for the same reason that vanity lights are wired in parallel—so that they all work.

Here's how to figure how many loads you can put on a transformer: A transformer comes with a VA rating. Typical is 20VA or 40VA. VA is volts multiplied by amps. This is the same math as that used to figure watts. So you can assume that VA and watts are the same thing. If you have a load that has only an amp rating, multiply that by 24V to get the VA. The total VA ratings and watts of all the loads cannot exceed the transformer VA rating. If they do, the transformer dies.

Code

In the interest of safety, there are lots of requirements for how electricity is installed in a building. The national code is called the National Electrical Code (NEC). Local jurisdictions have their own variations. The code is lengthy and detailed. One of the services an electrician gives is knowing code and local applications. To add to the complexity, its interpretation and enforcement can vary considerably from inspector to inspector. When there's new construction or a remodel, the work must be inspected and determined "up to code."

In the end though, the code provides us with a national

standard of not only safety, but also convenience. Interesting examples of what the code provides for us are:

- There must be at least one electrical outlet on every wall.

- If a room doesn't have an overhead light, there must be a switched outlet to plug a lamp into.

- The kitchen must have a minimum of two heavier-duty, 20-amp circuits for appliances.

- GFCI outlets are required in areas close to water, such as the bathroom and kitchen, and outdoors.

Electrical devices often show the letters "UL." UL stands for Underwriter's Laboratories, which is a group that tests for safety standards.

240-volt circuits

Three lines come into the service panel from the electric utility. Two are 120V and one is neutral. A 240V circuit is created for heavy appliances by using both of the 120V lines.

Three-phase electricity

Large buildings sometimes use three-phase electricity because it's much more efficient than single-phase. But because it requires three transformers or a special line, the utility charges much more for it. It's not used residentially.

A phase is the difference between positive and negative

in alternating current (AC) electricity. Ordinary single-phase electricity shows a simple sine wave on the lab instrument called an oscilloscope. Three-phase electricity shows three sine waves. There is no two-phase electricity.

DC circuits

DC means direct current as opposed to alternating current (AC). DC appears on an oscilloscope as a straight line rather than as a sine wave. Batteries provide DC. Also, many plug-in appliances, including computers, internally change AC into DC. DC motors are often used in heating and air conditioning equipment. Advantages of DC are less noise, hum, and interference.

Conclusion

Now you understand house wiring. You aren't ready to fix your own wiring—you may need an electrician. And you can't design and install your own circuits. But you do know how electricity works, and that's more than most people know.

You know that wiring is made up of circuits, and that every circuit consists of a power supply, at least one switch and at least one load. You know that wiring everything up like beads on a string doesn't work. That's called series wiring. You know that house wiring is done in parallel. You can picture the wiring inside you walls. You can look at two outlets and "see" that electricity goes through and around an outlet at the same time.

And you know how to perform basic tests on an outlet or a switch. If you want to do more, you can either get an electrician or learn a lot more yourself. To learn more, there are lots of classes. There are also many wonderful full-color, step-by-step instruction books available at bookstores, online, and free at the public library. See "References" on the next page for a list.

References

Allen, Benjamin, ed. *Step-By-Step Wiring*. Des Moines: Better Homes & Gardens.

Black & Decker. *The Complete Guide to Home Wiring: A Comprehensive Manual from Basic Repairs to Advanced Projects*. Minnetonka, MN: Creative Publishing International, Inc.

Cauldwell, Rex. *Wiring a House*. Newtown, CT: The Taunton Press.

Erickson, Larry, ed. *All About Wiring Basics*. Des Moines: Ortho Books.

Fucci, Joseph, ed. *Wiring Basic and Advanced Projects*. Upper Saddle River, NJ: Creative Homeowner Books.

Richter, H. P. and W. Creighton Schwin. *Practical Electrical Wiring*. New York: McGraw-Hill.

Sidey, Ken, ed. *Basic Wiring*. Des Moines: Stanley Books.

Index

Order Information .

To order copies of
Quick & Basic House Wiring,
Quick & Basic Electricity,
Quick & Basic Hydronic Controls,
Quick & Basic Troubleshooting

Mail this form to:
P.I.G Press
759 E. Philips Drive S.
Littleton, CO 80122-2673

Enclose $20.00 US funds for each book, plus $3.95 for shipping
and handling.

Number of books requested: _____

Total enclosed: _____

Your Name: _____

Your Address (include zip code): _____

Questions? Call P.I.G. Press at:
303-795-2679
or Fax 303-795-9350

Thank you!